ROBOTS

TOP THAT! Kids™

Copyright © 2003 Top That! Publishing plc
Top That! USA, 27023 McBean Parkway, #408 Valencia, CA 91355
Top That! USA is a Registered Trademark of Top That! Publishing plc
www.topthatpublishing.com

Contents

Introduction	3
What is a Robot?	4
History	6
Robot Functions	8
Big Screen Robots	10
Small Screen Robots	12
How do Robots Work?	14
Robot Mechanisms	16
Robot Control Systems	18
Robots in Industry	20
Explorers	22
Bomb Disposal	24
Security	26
Space Missions	28
Mars Exploration Rovers	30
Robots at Home	32
Robots in Medicine	34
Robot Fun	36
Robot Toys	38
Artificial Intelligence (AI)	40
Nanobots	42
Future Robots	44
Glossary	46
Find Out More	48

Introduction

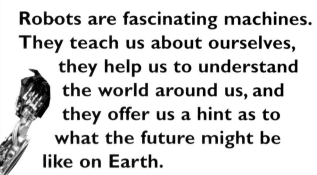

Robots are fascinating machines. They teach us about ourselves, they help us to understand the world around us, and they offer us a hint as to what the future might be like on Earth.

Robot Body
Robots are usually modeled on human bodies. But instead of eyes, a robot has sensors, instead of muscles, hundreds of tiny motors, and instead of a brain, a computer. By learning about robots we discover how complicated our own bodies are.

Our World
Robots help us to understand the world around us because they can go to places too dangerous for humans. Robots can travel deep under the sea, into volcanoes and land on planets in space. They bring back valuable information about these places, giving us a better understanding of the world around us.

Future
As technology progresses robots will be able to do more and more. Many people believe that one day they will even be able to think for themselves.

What is a Robot?

The word "robot" did not exist in the English language until the twentieth century, when it was used in a play.

The Word

Czech playwright Karel Capek wrote *R.U.R. (Rossum's Universal Robots)* in 1921. The play tells the story of a

Karel Capek.

world in which all the work is done by machines. These machines are called "robota," the Czech word for forced labor. Although Capek wrote the work in Czech it proved so popular that the word "robot" passed into the English language.

Definition

A robot is really any automatically-operated machine that replaces human effort. However, a robot does not have to look like a human being and nor does it have to perform its work in the same way as a human.

Robot Principles

In the 1940s famous sci-fi writer Isaac Asimov defined the operating principles of robots as follows: robots must never harm human beings; robots must follow instructions from humans as long as they don't break the first rule, and robots must protect themselves without breaking the other rules.

Asimo android made by Honda.

Android

An android is simply a robot designed to look like a human being. They are also sometimes known as humanoids.

What is a Robot?

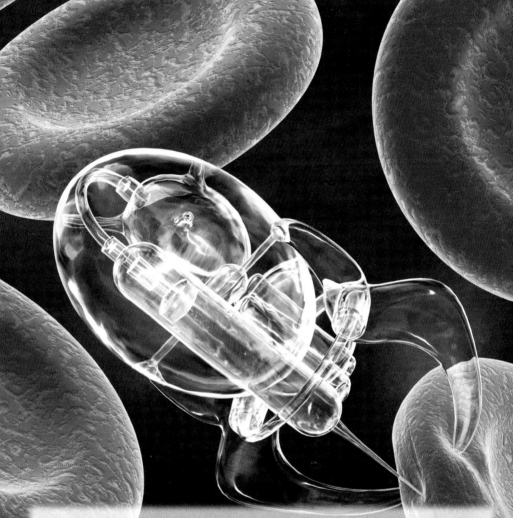

Nanobot
Nanobots (or nanorobots) are the smallest robots around. Their dimensions are measured in millionths of millimeters. There are two types of nanobot, autonomous and insect. Autonomous robots have computers on board but insect robots are controlled by a central computer. Nanobots are in the early stages of development but it is hoped that they will be used in medicine.

History

Although robots seem like a very modern idea, they are actually much older than you might think.

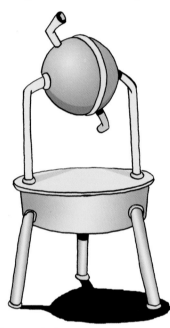

A water clock.

First Robot

The first invention to have a robotic function was probably the water clock built by ancient Greek inventor Ctesibius of Alexandria in around 250 BC. This used a siphon to recycle the water and tell the time. As the water clock worked without requiring any human effort it could be classed as a robot.

Industrial Revolution

During the eighteenth and nineteenth centuries the demand from industry for primitive robots rapidly increased. The steam engine was partly controlled by a "governor," a device operated by rotating weights. The weights on the governor determined the level of flow of steam to the engine. The steam engine's replacement, the internal combustion engine, used pistons that repositioned themselves after each cycle.

Twentieth Century

Robots have really come into their own in the last hundred years. While earlier inventions were able to carry out some tasks previously undertaken by humans, these machines still required a high level of human involvement. Today machines that require virtually no human involvement are commonplace. They are now seen in factories, homes, hospitals, space, and even in the occasional restaurant.

A steam train.

History

First Working Robot

The first working robot in the United Kingdom was made in 1928. Designed by Captain Rickards and AH Renfell, it was an electromechanical robot with an electric motor, electromagnets, pulleys, and wheels.

Around the World

Robots are now found in many countries but Japan has more than any other. The second most prolific country for robots is the United States.

T3

T3, the first commercially available minicomputer-controlled industrial robot, was designed by Richard Hohn in 1973. This machine used hydraulics to lift weights of 100 lb.

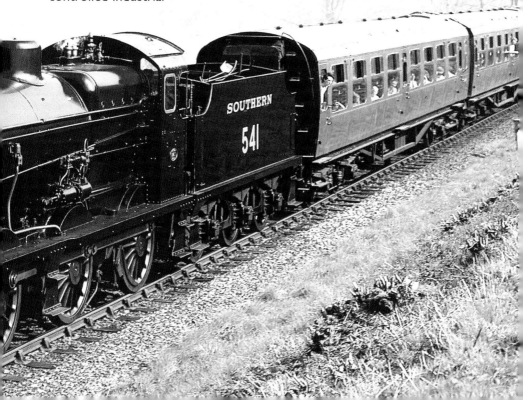

Robot Functions

Robots have different purposes—some are used to do really repetitive jobs that humans find boring while others are used for dangerous work.

Machine Parts

Robotic arms are often used in factories in order to paint machine parts or to assemble electronic circuits. Of course, although robots are expensive to build and design, they do not have to be paid which means that, in the long run, it works out cheaper to use a robot than to employ a human to do the same job.

An automated factory assembly line.

A remote-controlled bomb disposal robot.

Bomb Disposal

Before robots, bomb disposal was a dangerous business with only the bravest of soldiers prepared to risk their lives to defuse a bomb. But, today remotely-controlled robots can do the job and, if the attempt goes wrong, only the robot is damaged.

Robot Functions

Medicine
Some of the tiniest robots are used in medicine. They can be so small that they are able to travel through our bloodstream. The advantages of this are threefold: once the robot is successfully programmed there is little prospect of error; it can reach parts of the body awkward for a surgeon, and it cannot transmit disease to the patient.

Space Exploration
Although people have successfully landed on the moon, it has not so far been possible to get a human being onto any of the planets in our solar system. However, NASA scientists have sent robots to Mars. These robots have been used to help us learn about the geology of the "red planet."

A Mars explorer robot.

Domestic
There are many robots available for use around the home today. The problem is they are often expensive and so only the richest people can afford them. Robotic lawnmowers, vacuum cleaners and even home security robots are now available.

A robotic vacuum cleaner.

Big Screen Robots

Robots have always caught the imagination, and have featured in some of the most successful films of all time.

The Matrix
In this film, the world has secretly been taken over by robots living 200 years in the future. Without even knowing it, humans are used by the robots as a fuel source.

Artificial Intelligence: AI
Steven Spielberg directed this film, which stars Haley Joel Osment as David, the first robot to have feelings. When his mother rejects David in favor of her human child he struggles to regain her love.

Blade Runner
In this cult classic, Harrison Ford stars as a man charged with hunting down artificially created humans who have hijacked a space ship.

Bicentennial Man.

Bicentennial Man
Bicentennial Man began life as a short story by Isaac Asimov before being made into a film. It tells the tale of a robot's 200-year journey to become human. At first, just a domestic robot intended to perform household chores, bicentennial man gradually gains human emotions.

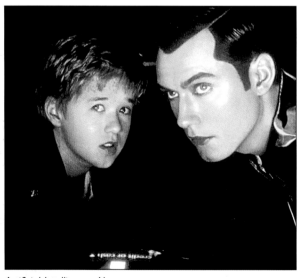

Artificial Intelligence: AI.

Big Screen Robots

Westworld
This classic 1970s sci-fi film tells the story of a futuristic amusement park where robots are employed to play whatever roles the customer wants. But when they are asked to perform a wild west adventure one of the robots begins to behave independently and starts shooting at the customers.

Star Wars
Star Wars, one of the most popular films of all time, features the robots R2D2 and C3P0. Unlike many robots on film, these androids are good, not evil. They help the film's hero Luke Skywalker, in his battles against Darth Vader.

Terminator
A human-looking robot, played by Arnold Schwarzenegger, is sent from the future to kill Sarah Connor, so that her son will not be born. The son is destined to lead the humans in a war against robots but stopping his birth will make this impossible.

Short Circuit
This comedy film tells the story of an experimental robot called Number 5. After Number 5 is electrocuted, it develops self-awareness and a fear of being re-programmed. It therefore tries to escape its owners.

Terminator.

Short Circuit.

Small Screen Robots

While perhaps not so popular as their big-screen rivals, the robots of television have also enchanted their audience.

Dr. Who
Quite a few robots appeared in this long-running British television show. The most popular were the daleks, who looked like shuttlecocks on wheels. Warrior robots, who had difficulty negotiating stairs, they were famous for their shout of "Exterminate."

A Dalek.

K-Nine
K-Nine was a dog-like robot given to Dr. Who as a present. Like most dogs, it could wag its tail but unlike other dogs, it could speak, too.

A Cyberman.

Cybermen
The cybermen of Dr. Who started off as humans but eventually became immortal by replacing their body parts with metal or plastic bionic parts.

Small Screen Robots

Marvin the Paranoid Android

Marvin the Paranoid Android starred in the cult series, *The Hitchhiker's Guide to the Galaxy*. He accompanied a human and an alien on their journey through space.

Six Million Dollar Man

Very popular in the 1970s, *The Six Million Dollar Man* starred Lee Majors as Steve Austin, a pilot very seriously injured in a flying accident. His broken limbs and damaged eye were replaced by artificial ones that gave him great strength and extraordinarily powerful vision.

Lee Majors.

Star Trek

Data from *Star Trek: The Next Generation* is a curious robot. He longs to feel the emotions of humans but when offered the chance to become human, refused. He has now been given an "emotions chip" but this nearly caused his system to overload.

Inspector Gadget

Inspector Gadget was the hero of his own cartoon. A police officer equipped with bionic powers, his arch-enemies were Dr. Claw and Madcat. It was much to his advantage that he could acquire legs with tremendous spring or a propeller for his hat at a moment's notice.

Data from Star Trek.

How do Robots Work?

All robots use moving parts, power to make the parts move and something to control the moving parts.

Humanoids
Whether or not the robot is an android it will, like a human, have a body structure, "muscles" to move the body structure and a sensory system to collect information about the body and the surrounding environment. The major difference is that, unlike humans, robots cannot think for themselves.

Robots have a body structure.

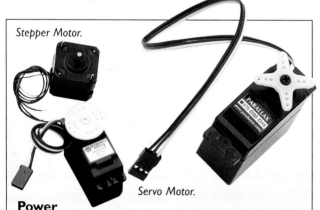
Stepper Motor.
Servo Motor.

Power
In order to do anything at all robots need power. This will either come from a battery or from a mains electricity source. Robots can be driven by AC, DC, stepper, and servo motors.

Computer
All robots have a computer for a brain. The robot can only act as commanded by the computer. If the computer has only been programmed to give the robot one command, then that will be the only order the robot can follow.

How do Robots Work?

Movable Body
Although some robots are more sophisticated than others, all can move in some way. Some use motorized wheels to move the body as a single unit while others have lots of different sections which can move. These movable sections are usually made of either plastic or metal.

Hands
The most sophisticated robot's hands, such as the robonaut's, are able to grasp and manipulate all sorts of objects. In this way, they replicate the human hand.

Housing
A robot is full of sensitive electronics which need to be protected. Fortunately, unlike a human, a robot's body does not require a heart, lungs, or, indeed, a kidney. This means that the robot's body can be used to house and protect the vital electronics.

Robot Mechanisms

Most robots, but particularly androids, are modeled on humans. Designed to have some, or all, of the same capabilities, their capacity is limited only by the imaginations of their inventors.

Hands
Hands are one of the most difficult things for robot makers to design. Made of metal or plastic segments, each segment is powered by a tiny motor. From the point of view of control, the hand is broken down into two sections, one for manipulation and one for gripping.

The forearm and wrist can be made strong enough to carry objects too heavy for a human hand.

Arms
The arms are packed full with joints and avionics—so much so that they have greater maneuverability than a human arm. The endoskeleton of the arm holds sensors and fail-safe brakes.

Head
As with a human, seeing and hearing devices are found on the head. The difference is that the robot's head can turn 360° independently.

This robot arm is pouring a cup of tea.

Robot Mechanisms

Actuators

Robots use a device called an actuator to spin wheels and pivot jointed segments. Some robots use electric motors and solenoids (a coil of wire through which an electrical current is passed) as actuators, some use a hydraulic system, and others use a pneumatic system. The actuators are connected to an electrical circuit and are controlled by the robot's computer.

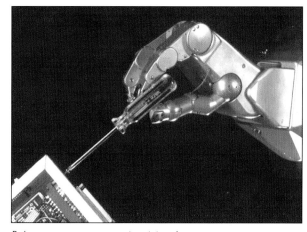

Robots use actuators to pivot jointed segments.

Sensors

A robot's ability to move without crashing into things is created by attaching an LED (light-emitting diode) to the slotted wheels of the joints. The LED shines a beam of light through the slots to a light sensor on the other side of the wheel. When the robot moves a particular joint, the slotted wheel turns. The slots break the light beam as the wheel spins. The light sensor understands the pattern of flashing light and transmits the data to the robot's computer. The computer can then tell exactly how far the joint has turned.

K9 Rover.

17

Robot Control Systems

Although robots may seem to act independently, they are in fact controlled by humans, who write the computer programs which make them function.

Switches

At the most basic level, robots are controlled by a system of computer controls and switches. A relatively simple computer program can be written to instruct the robot. For the robot to understand, the instructions must be along the lines of "If this happens, then do this." For example, a robot could be made to understand the instruction "If sensor A is activated, then turn off motor B."

Voice Activation

Some robots are trained to recognize their master's voice and respond to their commands. This technology is not 100% efficient as it takes time for the robot to distinguish the voice from others and to pick it out if other people are speaking at the same time.

Internet

Some robots can be controlled by Internet communication. The human controller can send messages in e-mail form to the robot which has been programmed to understand and respond to the instructions.

Robot Control Systems

Moving Parts

Control of the moving parts of the robot is achieved through the robot's "muscles"— really a system of gears and chains. There will be gears controlling individual areas of the robot as well as a gear to enforce movement of the whole body.

A toddler robot which uses gears and chains to enable it to move.

Remote Control

Robots can also be operated by remote control. Sensors on the robot will respond to the movements of a joystick or commands entered in a keypad.

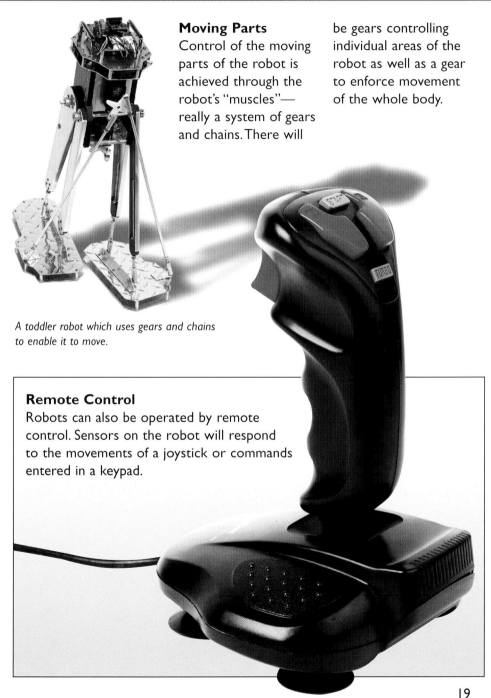

Robots in Industry

Robots are now a common sight in industry, freeing up humans for more complex tasks.

Efficiency
For a robot to be employed in industry it must be more efficient than a human and less costly to run. Some people, known as Luddites, believe that this progress in technology takes jobs away from people. However, when robots take over they create new jobs as now people are needed to build and look after the machines.

Painting
A robot which could properly handle a tool (for painting) was first used in Norway in 1966, and in the USA, robots were developed for spot-welding on assembly lines. Since then, there has been a continual evolution towards robots of greater precision, such as the Japanese selective compliance assembly robot arm (SCARA).

Welding
The most popular application for robots is welding. Using robots improves safety, as no human will be hurt, efficiency, as a robot never misses a welding spot and costs, as a robot doesn't have to be paid. Around a quarter of the robots used in industry are used for welding work.

Materials Handling
Another industry in which robots are popular is materials handling. The Nimbl H-series Scara, with its ability to work quickly and efficiently, is one example of a robot well suited to picking up and placing materials.

A welding robot in action.

Robots in Industry

Motoman
The Motoman UPJ is a vertically articulated robot capable of moving weights of 8 pounds at a time. As such, it is ideal for handling small objects. It can be used for assembling, testing, and general handling applications.

Computers
Manufacturing robots are vital to the computer industry. A precise hand is needed to put together a microchip, the tiny electronic circuit at the heart of today's computers.

The PUMA
The PUMA (Programmable Universal Machine for Assembly) is the most widely used industrial robot. Designed in Switzerland in the 1970s, this robot is commonly found in laboratories and automated assembly lines. More cars are now made by robots than by humans.

Programmable Universal Machine for Assembly: PUMA.

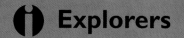 Explorers

Although there are plenty of famous human explorers, none have been as intrepid as their robot equivalents.

Slocum Seaglider.

Seaglider

Autonomous Underwater Vehicles (AUVs), such as the Slocum Seaglider, are used for deep sea research under the ocean. The Seaglider doesn't use motors or fuel to move through water. Instead its variable buoyancy allows it to sink or swim at will. Before being launched the Seaglider is programmed with the coordinates of which part of the ocean it is to research. A small electric pump then transfers enough oil into the Seaglider to make it heavy and dense enough to sink.

Up Again

Once the Seaglider reaches a predetermined depth, it will measure such things as the water temperature and its salt content. When the research is complete the oil is transferred back out of the Seaglider so that it can rise to the surface.

Explorers

Dante II.

Dante II
Dante II was launched in 1994. This six-legged robot was programmed to explore the active volcano of Mount Spurr in Alaska. It was slightly more successful than Dante I, which fell into the fires of the volcano after its cable broke not far from the surface. Dante II was able to gain some gas and water samples and send some video footage before its mission also ended in disaster. After being damaged by a falling boulder it then fell into the volcano. An attempt to rescue it by helicopter failed when the tether holding the robot broke.

Nomad
This robot crossed the Atacama Desert in Chile in order to find out about a landscape similar to the surfaces of the moon and Mars. The aim of the trip was to make sure that when a mission went into space the robot would be able to negotiate the terrain and return with valuable data.

Nomad.

Bomb Disposal

If a terrorist plants a bomb, there's a good chance that lives will be lost. But if a robot can be used to deactivate the device even soldiers will not be at risk.

Design

There isn't time for a bomb disposal robot to be built after the device has been discovered. Designers have therefore had to do a great deal of research to make sure their robots are well suited to the task.

Maneuverability

Bombs are rarely left in open spaces—in fact, 90% of the time they are left in buildings. The robot, therefore, needs to be able to negotiate stairs, turn sharp corners, and may even need to open doors.

Timing

Most bombs are intended to go off when it is dark. Robots must therefore be fitted with lighting or imaging systems so that the robot's controllers can see what's going on.

The effects of a bomb can be devastating.

Approach to Threat

The robot enables the bomb to be disposed at little or no risk to human life. Even if the approach to the bomb is booby-trapped the bomb disposal officer remains at a safe distance from the threat. This safety is vital because it helps the officer to remain calm in what would otherwise be a very stressful situation.

A remote-controlled bomb disposal robot.

24

Bomb Disposal

First Bomb Disposal Robots

The first robots, intended as a response to improvised explosive devices, were built in the early 1970s. The first ones were so basic they were known as Wheelbarrows. They could do little more than carry the bomb to a safer place. It wasn't until the fourth version of the model, the Wheelbarrow Mark 4, that it was possible to view suspicious objects remotely via a closed circuit camera.

The Andros Mark 5

The evolution of bomb disposal robots has now reached the Andros Mark 5. This is a robot built to operate in any weather condition. It can identify and dispose of explosive devices, it can X-ray suspicious devices to assess their danger level, it has audio and video facilities for surveillance work, and it has audio communication facilities which can be used in hostage situations. It even has a shotgun should it get into an argument.

Power

The Andros Mark 5 weighs 600 pounds. It can lift 70 pound weights, pull 400 pound weights, climb 45° slopes or stairs, cross 3 foot ditches and climb 20 inch high ledges.

One of the latest Wheelbarrow robots.

Security

Robots are great security providers, not just for the military but also for industry and the home.

Cyberguard

Cyberguard, a robot manufactured by Cybermotion,

Cyberguard can detect smells.

is intended to protect factories from crime. Unlike a human security guard, there's no chance of it becoming tired—its concentration level remains the same 24 hours a day. Cyberguard travels around factory floors searching for anything that looks suspicious. Its findings are then relayed to a central control by a real-time video link. What's more, Cyberguard is programmed to detect suspicious smells as well.

MARON-1

MARON-1 operates as a home security guard. As well as being able to take pictures and relay them to the owner's mobile phone, this robot can look after the air conditioning system. The robot moves about on two wheels and has sensors to prevent it bumping into household objects. Another aspect of MARON-1 is that it can be controlled remotely by the owner.

Banryu

Banryu is another home security guard. Designed to look like a dog it is 36 in. long, 32 in. tall and 28 in. wide. Weighing 90 pounds, this dog can

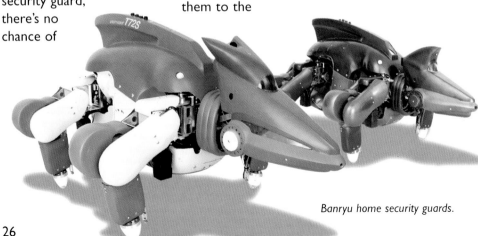

Banryu home security guards.

Security

detect smoke and intruders. When the owner is at home, it can be set to pet mode, when it will respond to orders to beg or sit.

SARGE.

SARGE
As you might expect, SARGE (Surveillance and Reconnaissance Ground Equipment) is a military robot. Developed for the US Marines, SARGE is sent into the field first. It helps the soldiers to decide what their future needs on the battlefield will be, so minimizing their personal risk.

AROD
Attempts were made to carry out surveillance from the air with an AROD (Airborne Remotely-Operated Device). Small enough to be carried by one person they could be powered from a tether on the ground. Although ARODs proved successful in spying, they proved to be unstable in flight and were discontinued.

AROD.

Space Missions

Although we tend to think of robots as highly efficient machines, their missions to space have met with varying degrees of success.

Viking

The 1975 Viking mission to Mars transmitted tremendously valuable information about the planet's surface. The Viking mission dropped a robot on the planet's surface, which then sent images to scientists at NASA. The mission showed that the planet appeared to be divided into two main regions with plains in the north and cratered highlands in the south. It was also discovered that there could be 90°F variations in temperature in a single day.

Mars Observer

The Mars Observer mission of 1992 ended in disappointment when contact was lost with the spacecraft

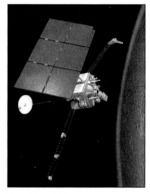

The Mars Observer.

before it had begun its orbit of the planet. This was an expensive loss as the Observer contained such sophisticated instruments as a high-resolution camera,

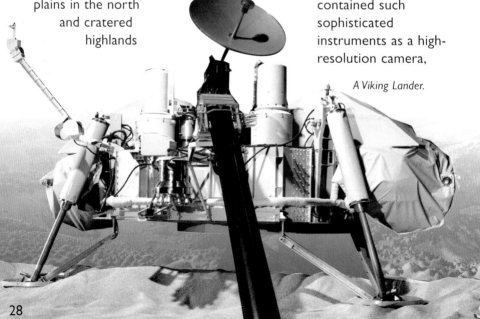

A Viking Lander.

Space Missions

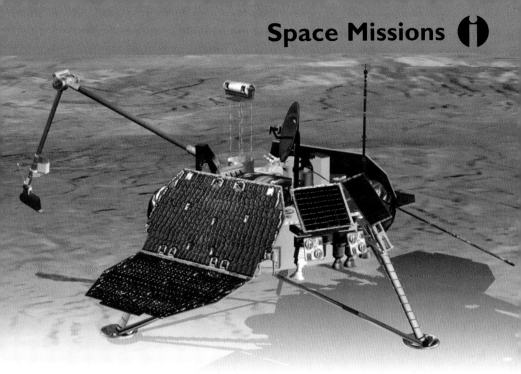

The Mars Polar Lander.

a thermal emission spectrometer, a laser altimeter, a magnetometer, a pressure modulator infrared radiometer, and a gamma ray spectrometer.

Mars Polar Lander

The Mars Polar Lander mission of 1999 also ended in disappointment as contact was lost with the spacecraft before it could provide any useful information. The aim of the mission was to land a spacecraft near the planet's south polar cap and dig for water with a robotic arm.

Venus Vega

In 1985, the Venus Vega recorded temperatures, pressures, wind speeds, and cloud particle properties of Venus. Not designed to land on the planet, the Venus Vega was an aerobot, a robot intended to float in a planet's atmosphere.

Aerobot Data

Planetary aerobots must be able to operate with little or no communication from Earth. Aerobots can work out where they are, control their altitude, and gain scientific information. Aerobots are able to float because of the buoyancy of the gas they carry. They control their altitude by releasing ballast or gas.

Mars Exploration Rovers

The Mars Exploration Rovers are among the latest space exploration robots to be built.

NASA's Mars 2003 Rover.

Bad Atmosphere

There is so much carbon dioxide in the atmosphere of Mars that it has not yet been possible for an astronaut to land on the planet. However, it has been possible to send robot researchers to the "red planet." But none have been as sophisticated as the Mars Exploration Rovers.

Equipment

Mars Exploration Rovers are equipped with a mobility system so that they can move, a tall mast for a good view of their environment, a robotic arm so that they can position tools against interesting rocks, solar panels for power, insulation to protect the electronics, instruments for sensing the environment, and tools to prepare the rocks for examination.

The Mobility System

Each wheel has an independent motor. The two front and two rear wheels have individual steering motors, enabling the robot to turn. The robot also has a four-wheel drive option so that it can swerve. The robot's suspension system

Mars Exploration Rovers

enables it to negotiate slopes of 45° and to drive through small holes. They can cover 300 feet in one day.

A tiny Nanorover.

The Robotic Arm

The robotic arm comes complete with arm, shoulder, elbow, and wrist. This arm enables scientists to place four instruments exactly where they want to against a rock.

Insulation

Insulation is provided by the warm electronics box, a heated container. Without this heat, the variations in the temperature at the Martian surface would cause the valuable sensors to break.

Solar Panels

The solar panels generate enough power to recharge the robot's batteries, which drive the robot's instruments.

Rock Examination

The robot is equipped with a rock abrasion tool (RAT).

A K9 Rover.

The RAT grinds a hole into the surface of the rock so that scientists can see both the inside and the outside of the rock. The RAT's rotating brush sweeps away the dust created by the grinding mechanism.

Rocky 7 Rover.

Robots at Home

Robots are not just in demand in the workplace, they also have much to offer us at home.

Roomba– vacuum cleaner.

Roomba
Roomba is a robotic vacuum cleaner. It uses intelligent navigation technology to clean nearly all household surfaces without human direction. It even knows when it's finished so it can turn itself off. You'll still have to put it away though.

Robomower
Working on similar principles to the robotic vacuum cleaner, the robomower cuts the grass all by itself. All you need to do is take it to the grass and press the start button. Once you've marked out the perimeter with wire, it will even recognize the edges of your lawn.

Dr. Robot
Dr. Robot has a dual function: he can be either a home security guard, or an entertainment system. As well as being able to walk and talk, Dr. Robot can detect an intruder in the home and then contact the owner on their cell phone to warn them that there is trouble in store.

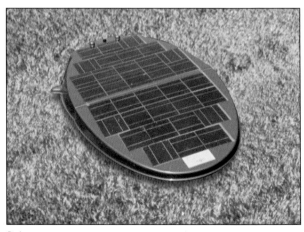

Robomower.

Robots at Home

Teksta.

Cyber Pets
Of course no home is complete without a pet, whether furry or robotic. The Teksta puppy is a robotic toy ideal for young children. A voice-activated dog, Teksta will walk, talk, and wag its tail. It will also learn about its environment.

Muy Loco
Another cyber pet, this one has a South American feel to it. Activating the sensor on the back of Muy Loco's neck will make him do some Latin American dancing. What's more, Muy Loco sticks his tongue out on demand!

Chores
The demand for robots often comes from a human desire not to have to do tedious, repetitive jobs. Many people are just waiting for robots to do the ironing, dusting, and clean the bath. The first person to build a complete domestic servant robot will no doubt become very rich indeed.

The Dyson DC06 robotic vacuum cleaner.

Robots in Medicine

One area in which robots are beginning to prove their value is medicine. This is just one of the professions where robots are showing they can do more than boring, repetitive work.

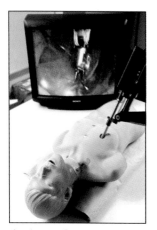

A robot performing an operation.

Benefits

Robots are unlikely to make mistakes because, once they have been successfully programmed, there is no possibility of human error and little chance of infection. They don't get tired, their accuracy is perfect, and they can keep going for 24 hours a day. Also, the smallest ones can get to places that surgeons and doctors can't easily reach.

Brain Surgery

Robots help in brain surgery by guiding the tip of the surgical instrument into the brain through a small hole drilled in the skull. Although this is very helpful as it minimizes the area to be operated on, it wouldn't be of much use if the surgeon couldn't see the area of the brain to be operated on. This problem was solved by programming the robot to send back to the surgeon a three-dimensional image of the operation site.

Telepresence

In 2001, an operation to remove an elderly woman's gall bladder took place without the surgeon ever entering the room. The surgeon operated on the three-dimensional image and his actions were then replicated by the robot in the operating room. A high-speed

A robot-guided surgical instrument.

Robots in Medicine

telecommunication link between the surgeon and the robot meant that the robot's actions took place 0.1 seconds after the surgeon's. In a 2002 operation using telepresence, a surgeon in Israel operated on a man in Italy.

Physiotherapy
A team at the Massachusetts Institute of Technology is developing a robot to help patients regain movement. The patient puts their hand on a robotic joystick which guides them through a series of movements. Unlike a human physiotherapist, the robot can guide the patient thousands of times without becoming tired.

In the Bloodstream
Robots have already been invented that are so small they can travel around our bloodstream. It is hoped that they will be able to repair cells and stop viruses and nasty bacteria in their tracks.

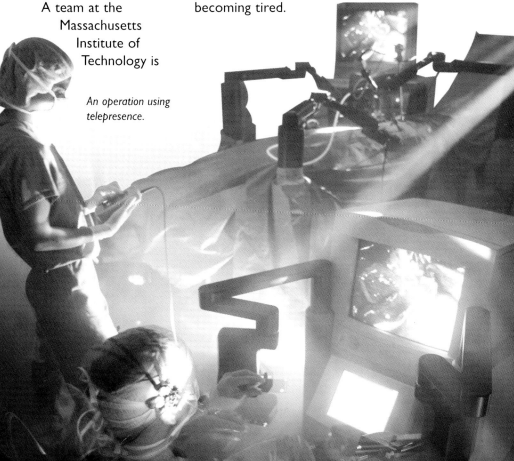
An operation using telepresence.

Robot Fun

Now let's look at some fun robots, both from history and from around the world!

Minerva
A humanoid robot called Minerva used to guide tourists on an hour-long tour of the National Museum of American History. Minerva greeted guests, commented on what they were wearing, and offered them information as she accompanied them on their tour of the museum. She also smiled when people asked questions and blasted her horn when objects blocked her path.

Robot Trolley
In Yo! Sushi, a restaurant in London, there is a robot waiter. While it can't take your order, it will bring your drinks to the table. It has sensors to prevent it bumping into things, and can detect people if they get in its way.

Chess Player
In the eighteenth century, Baron Wolfgang von Kempelen created a "robot" who was very good at chess and who he dressed as a Turkish gentleman. It was later discovered that the robot was actually operated by someone hiding under the table.

Minerva robot. Lemelson Center for the Study of Invention and Innovation.

Robot Fun

PaPeRo Robots.

PaPeRo
PaPeRo is a personal robot who likes to have fun. Designed to walk, talk, and deliver messages, he will wander around the room looking for something to do if he hasn't been given a task. If no one has anything for him to do, he takes a nap.

Elektra
In 1940, Elektra the robot was launched by the Westinghouse Electrical Corporation. Elektra could dance, count to ten, and smoke. What's more, he had a pet dog, Sparko, who could walk, stand on its hind legs and bark.

Dream Robot
In 2000, Sony launched a humanoid called the Sony Dream Robot. This robot can distinguish between colors, kick a ball into a net and, in a real advance for this type of robot, move the upper half of its body to counter movement in the bottom half.

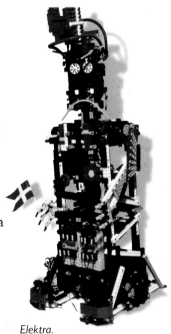

Elektra.

Robot Toys

Every now and then a robotic toy captures the imagination and children everywhere have to have one.

Transformers

Some of the most famous robots were Transformers, spin-off toys from an animated television series. In the series the Transformers were robot warriors from the planet Zobitron. They fought their battles against their enemies, the Decepticons, on Earth.

A Transformer toy that changes from a vehicle to a robot.

Sony AIBO

The Sony ERS-220 AIBO Entertainment Robot 220 is an entertainment robot. When put in autonomous mode, it will make its own decisions. Its built-in curiosity allows it to gain experience and so develop its character. This robot can also express happiness, sadness, fear, dislike, surprise, and anger.

Soccer Pro II

The Soccer Pro, a robot you can build, uses a wired controller to activate movement and ball control. It can run forwards, backwards, to the left and the right, and execute 360° turns.

Spider III Robot

This fun robot uses a light sensor beam to sense objects in its way. When an obstacle is detected it changes direction and starts on a new path.

Sony AIBO.

Robot Toys

Robot Kits
There are robot kits available, but you will need adult help to build one. Some of the best-known ones are the LEGO Mindstorms, the BOE Bot, and the ARobot.

ARobot.

LEGO Mindstorms
According to the manufacturer, a first-time user with a few computer skills will be able to build a simple LEGO robot in one hour. These robots can steer round obstacles, follow trails, and react to changes in light.

ARobot
The ARobot is a computer-controlled mobile robot that can be relatively easily assembled. It's very useful for robot students as it helps them to understand computer programming, motion control, sensors, path planning, and object avoidance.

BOE Bot
The BOE Bot has a sturdy platform for its servo motors and circuit board. There are also mounting holes and slots so that customised robotic equipment can be added. The robot can be programmed to follow a line, solve a maze, follow light and communicate with other robots.

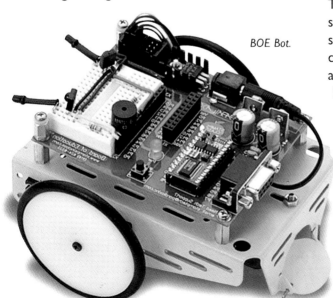

BOE Bot.

Artificial Intelligence (AI)

The idea that a machine could be created that could think for itself has always been popular in fiction, but few people have believed that it could become reality.

The Enigma Decoder.

Alan Turing

Alan Turing was a British scientist and mathematician who played a crucial role in World War Two by solving the secrets of Enigma, the machine used to encode German secret messages. A brilliant man, he came to suspect that anything created by the human brain must be a computable operation. It therefore stood to reason that a computer could be programmed with human intelligence.

Chess

Turing suggested that a computer could be programmed to outwit even the greatest human chess player. This was eventually shown to be true when the IBM computer Deep Blue defeated world chess champion Garry Kasparov in 1997. Although Deep Blue had won and shown that its capacity for computable operations was greater than Kasparov's brain, unlike Kasparov its thought processes had not been affected by emotions during the match.

Garry Kasparov.

Speech Recognition

Apart from chess, AI has had some success in speech recognition. It is now possible to program a computer or robot to recognize its owner's voice and respond to some simple instructions.

Artificial Intelligence (AI)

Impossible?

Today it is believed that an artificial intelligence machine that entirely replicates a human's intelligence will be very difficult to create. This is because a computer can only be programmed to make decisions such as "if this happens, then do that." But for a machine to really replicate human intelligence it would have to have the same range of knowledge as a human and also be affected by emotions, something machines don't have.

Kismet

Today, scientists at the Massachusetts Institute of Technology (MIT) are developing a robot called Kismet. This robot will be able to develop social skills by learning from its environment. By picking up on facial expression, body posture, gesture, gaze direction, and voice of humans it comes into contact with, Kismet will develop its own "personality."

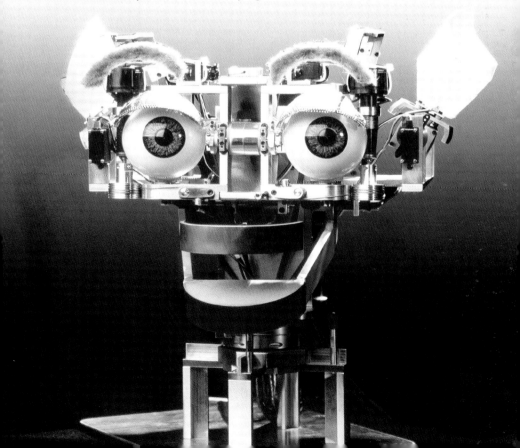

Kismet.

Nanobots

One of the most exciting developments in robotics is nanobots, robots so small it's difficult even to imagine them.

Young Technology
Nanotechnology is an experimental technology, and little is known about its full potential. Its aim is to use the smallest elements of a material, its atoms and molecules, as the parts for minuscule machines.

Purpose
A nanobot is a tiny nanomachine designed to perform a specific task or tasks repeatedly and with precision. Of course, its size means that it can negotiate the tiniest nook or cranny.

Size
The dimensions of nanobots are measured in nanometers. A nanometer is a millionth of a millimeter.

Two Groups
There are two types of nanobot, autonomous and insect. Autonomous nanobots have an on-board computer, but an insect nanorobot does not. Instead it is one of many, identical units all controlled by a single, central computer.

Scanning Electron Microscope
The scanning electron microscope is crucial to the development of nanotechnology.

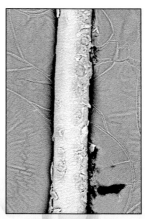

A human hair magnified under a scanning electron microscope.

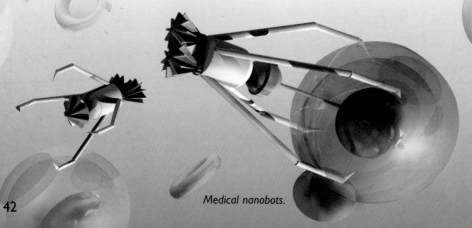

Medical nanobots.

Nanobots

This microscope is powerful enough to produce three-dimensional images magnified 200,000 times. This machine can also drill holes the size of a nanometer.

How?
A minute electrical impulse could drive the nanobot and the bot's limbs could be made from a copper molecule. This is not as unrealistic as it might sound as in 1991, a copper molecule 30,000 times as strong as steel and 50,000 times thinner than a human hair was discovered.

Why?
If nanobots can be developed successfully, a huge leap in the ability to heal the sick may follow. It is not beyond the realms of possibility that self-replicating insect nanorobots might one day act as a vaccine against disease. The nanobots would work by searching for and then destroying specific bacteria, fungi, or viruses. Nanobots also have potential applications in the assembly of small-scale, sophisticated systems. They might function at the atomic level to build devices, machines, or circuits one particle at a time.

A scanning electron microscope.

Future Robots

Although no one can say for sure what the future will be for robots, if they reach their potential the world will soon be a very different place.

Spider-bots
This high-tech critter may, one day, chart the terrain on other planets or explore smaller bodies such as comets, asteroids or the Moon. Spider-bots may also help with maintenance and repairs on the International Space Station. On Earth, they could fill in for humans by investigating hazardous materials or taking soil samples on farms. They have feeler-like antennas, which help to detect various obstacles. First prototypes have been built to a size able to fit in the palm of your hand. Future versions could be one tenth of that size, equipped with cameras that pan and survey its surroundings.

Medicine
If robots are performing operations, people will be attended to more quickly as robots can work 24 hours a day. There will also be a higher success rate as there will be no possibility of human error. What's more, using nanotechnology, it may soon be possible to insert a nanobot into the human body that will travel around the bloodstream killing nasty bacteria and viruses.

A Spider-bot prototype.

Future Robots

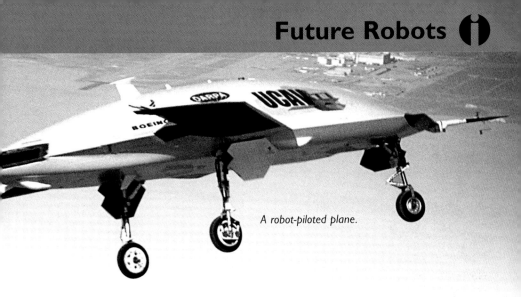

A robot-piloted plane.

Artificial Intelligence

All robot scientists dream of creating an intelligent robot. If this were to happen, robots could do virtually all jobs, not just the ones that require fairly simple decisions to be made. But the idea of artificial intelligence raises some difficult questions. If robots were built that were so much cleverer than human beings, would they rule us? And, what would we do all day if robots were doing all the work? Would we have fun or would we be bored?

Science and Technology

Where robots are really useful is in doing things that would be dangerous to humans, such as bomb disposal and space exploration. Preventing danger is an important aim for many robot scientists so perhaps robot soldiers, robot fire fighters and robot police officers will be one of the next developments.

A fire-fighting robot.

45

Glossary

Actuator Device
used to spin the wheels and pivot the jointed segments of a robot.

Aerobot
robot which can float in space.

Altimeter
instrument which measures the height above sea level.

Android
a robot that looks like a human being.

Andros Mark 5 A
bomb disposal robot.

ARobot
a robot that can be built from kit.

AROD
an airborne remotely-operated device.

Artificial Intelligence
ability of a machine to copy intelligent human behaviour.

AUV
Autonomous Underwater Vehicle.

Avionics
electronic circuits.

Banyru
a home security robot.

Bionic
powered by a mixture of human and electrical equipment.

BOE Bot
a robot that can be built from kit.

Cyberguard
an industrial security robot.

Cyberpet
a pet robot.

Dante II Explorer
robot built to search volcanoes.

Deep Blue
Chess-playing computer.

Dr Robot
an entertainment and home security robot.

Elektra
primitive robot.

Endoskeleton
internal skeleton.

Humanoid
another word for an android.

LED
light-emitting diode, used in sensors.

LEGO Mindstorms
robots that can be built from kit.

Luddite
person who is against technological progress.

MARON-1
a home security robot.

Mars Observer
1992 mission to find out more about the planet.

Mars Polar Lander
1999 mission to find out more about the planet.

Microchip
vital part of a computer.

Minerva
a robot tourist guide.

Motoman UPJ
a vertically articulated robot.

Muy Loco
a Latin American cyber pet.

Nanobot
a robot whose size is measured in nanometers.

Nanometer
a millionth of a millimeter.

Glossary

Nomad
an explorer robot.

NASA
America's space agency, the National Aeronautics and Space Administration.

P3
humanoid robot designed by Honda.

PaPeRo
a fun-loving toy robot.

PUMA
Programmable Universal Machine for Assembly robot.

RAT
Rock Abrasion Tool.

Robomower
a robot that cuts the grass.

Robonaut
space-traveling robot.

Roomba
robotic vacuum cleaner.

Rover Robot
used to investigate space.

SARGE
Surveillance and Reconnaissance Ground Equipment.

Scanning electron microscope powerful enough to produce three-dimensional images magnified 200,000 times.

SCARA
Selective Compliance Assembly Robot Arm.

Seaglider
an autonomous underwater vehicle.

Sensor
tool used by robot to sense objects in its way.

Shadow
robot with a wooden skeleton, upon which the muscles and equipment are mounted.

Soccer Pro II
soccer-playing robot.

Solenoid
coil of wire through which an electrical current is passed.

Sony Dream Robot
a fun-loving robot.

Sony ERS-220 AIBO
a fun-loving robot.

Sparko
elektra's robot dog.

Spectrometer
instrument for producing a spectrum.

Spider III
toy robot that looks like a spider.

T3
the first commercially available minicomputer-controlled industrial robot.

Telepresence
operating a robotic device remotely with a telecommunications device.

Transformer
toy robot.

Venus Vega
aerobot sent to research Venus. Viking NASA mission to Mars.

Wheelbarrow
bomb disposal robot.

47

Find Out More

Now that you've got a little taste of what robots are all about, you're sure to want to find out more!

Library
Why not go to your local library and see if they've got any books about robots? If they haven't, ask them to order one for you.

Internet
If you want to find out more about robots and space exploration, the best site to visit is www.nasa.gov. This is a great site, full of amazing information. www.thetech.org is another good site. It is a great robot resource with links to lots of other interesting sites. Always surf the internet under adult supervision.

Robot Clubs
If you're lucky, you might find there's a robot club near you. If not, why not ask your science teacher to start one up? You'll make some great friends and may soon be able to build your own robot. Some clubs even have competitions to encourage people to build robots with maneuverability, strength and the ability to climb.

Acknowledgments
Key: Top - t; middle - m; bottom - b; left - l; right - r; Science Photo Library - SPL.

Front cover: Sony Corporation. **Back cover:** (t) Parallax Inc.; (b) 2002 Sanyo Electric Co./ Ltd. **1:** 2002 Sanyo Electric Co./ Ltd. **2:** Stewart Cook/Rex Features. **3:** (l) Nasa; (m) David Gray/SPL; (r) 2002 Sanyo Electric Co./ Ltd. **4:** (l) Getty Images; (r) PGE/Rex Features. **5:** Coney/Jay/SPL. **6:** Meme Design. **7:** (t) Courtesy of NEC Corporation 2002; (b) Corel. **8:** (t) Maximilian Stock Ltd/SPL; (b) NJ/Rex Features. **9:** (t) Nasa; (b) Peter Menzel/Spl. **10/11:** Allstar pl. **12/13:** Allstar pl. **14:** (t) Nasa; (b) Parallax Inc. **15:** US Department of Energy/SPL. **16:** PMD/Rex Features. **17:** Nasa. **18:** Parallax Inc. **19:** (t) Parallaz Inc.; (b) TTAT. **20:** NDK/Rex Features. **21:** Nasa. **22:** (t) Webb Research Photo; (b) David Doubilet/SPL. **23:** (t) Hank Morgan/SPL; (b) Peter Menzel/SPL. **24:** (t) Sipa Press/Rex Features; (b) Labat Lanceau, Jerrican/SPL. **25:** Sutton-Hibbert/Rex Features. **26:** (t) Cybermotion; (b) 2002 Sanyo Electric Co./ Ltd. **27:** (t) Courtesy of Sandia National Laboratories, Intelligent Systems and Robotic Center (www.sandia.gov/isrc). (b) Space and Naval Warfare Systems Center, San Diego. **28/29:** Nasa. **30/31:** Nasa. **32:** (t) iRobot; (b) Philippe Plailly/Eurelios/SPL. **33:** (t) TTAT; (b) Dyson. **34:** (t) Hank Morgan/SPL; (b) Nasa. **35:** Nasa. **36:** Charles Townes/Smithsonian Institution. **37:** (t) Courtesy of NEC Corporation 2002; (b) Volker Steger/SPL. **38:** (t) TTAT; (b) Sony Corporation. **39:** Parallax Inc. **40:** (t) TTAT; (m) Pendzich/Rex Features. **41:** Sam Ogden/SPL. **42:** (t) TTAT; (b) Erik Viktor/SPL. **43:** Nasa. **44:** Nasa. **45:** (t) PMD/Rex Features; (b) Roy Garner/Rex Features. **48:** Peter Menzel/SPL.